COUNTDOWN TO SPACE

NASA PLANETARY SPACECRAFT

Galileo, Magellan, Pathfinder, and Voyager

Carmen Bredeson

Series Advisor:
John E. McLeaish
Chief, Public Information Office, retired,
NASA Johnson Space Center

E
Enslow Publishers, Inc.
40 Industrial Road
Box 398
Berkeley Heights, NJ 07922
USA

PO Box 38
Aldershot
Hants GU12 6BP
UK

http://www.enslow.com

Library of Congress Cataloging-in-Publication Data

Bredeson, Carmen.
NASA planetary spacecraft : Galileo, Magellan, Pathfinder, and Voyager / Carmen Bredeson.
p. cm. — (Countdown to space)
Includes bibliographical references and index.
Summary: Discusses the roles of several NASA spacecrafts which carried out probes during explorations of the planets.
ISBN 0-7660-1303-0
1. Space probes—Juvenile literature. 2. Planets—Exploration—Juvenile literature. [1. Space probes. 2. Planets—Exploration.] I. Title. II. Series.
TL796.3 .B74 2000
629.43'54—dc21
99-050639

Printed in the United States of America

10 9 8 7 6 5 4 3 2 1

To Our Readers: All Internet addresses in this book were active and appropriate when we went to press. Any comments or suggestions can be sent by e-mail to Comments@enslow.com or to the address on the back cover.

Photo Credits: Courtesy Lockheed Martin Corporation © 1999, p. 39; National Aeronautics and Space Administration, pp. 4, 8, 10, 12, 13, 15, 17, 18, 19, 22, 24, 25, 27, 28, 31, 33, 34, 35, 36, 37, 39; Photo: ESA, p. 20.

Cover Illustration: NASA (foreground); Raghvendra Sahai and John Trauger (JPL), the WFPC2 science team, NASA, and AURA/STScI (background).

CONTENTS

All these planet images have been taken by spacecraft. Included are (from top to bottom) images of Mercury, Venus, Earth (and Moon), Mars, Jupiter, Saturn, Uranus, and Neptune. Pluto is not shown, since no spacecraft has yet visited it.

1

Exploring the Planets

How do we know that the temperature on Venus is nearly 900°F or that the Martian sky is pink? And what about those beautiful rings around Saturn? Scientists tell us they are made up of chunks of ice, some the size of a bus! How do they know? Telescopes have given us a good look at our neighbors in space. Something has always stood in the way of even the most powerful telescopes though—Earth's atmosphere. The air that surrounds our planet keeps much of the light from space from ever reaching us. To find out more about the planets in our solar system, we must get close to them.

It was only after the development of the rocket during World War II that getting into space became possible. On October 4, 1957, the Soviet Union launched the first artificial satellite, *Sputnik 1*, into orbit around Earth. Since that time, hundreds of spacecraft have lifted off from Earth. Included in those missions have been many unmanned probes to the planets. Finally, people on Earth were able to get a close-up look at the mysterious worlds that share our Sun.

The nearest planet to the Sun is Mercury, followed by Venus, Earth, Mars, Jupiter, Saturn, Uranus, Neptune, and Pluto. NASA has sent probes to all of the planets except Pluto. There is not room in this book to examine all of these missions. Instead, we will take a look at the most recent spacecraft sent to Venus, Mars, Jupiter, and Saturn. We have learned much about our neighboring planets. Get ready for a fascinating look at what the probes found.

2

Magellan Orbits Venus

On August 10, 1990, scientists and engineers at NASA's Jet Propulsion Laboratory (JPL) in California watched nervously as the spacecraft *Magellan* began to do its work. When data started arriving from Venus, the scientists cheered and patted each other on the back. After completing one orbit of Venus, *Magellan* started on a second one. Then nothing! No more data arrived. *Magellan* project scientist Stephen Saunders exclaimed, "We knew the spacecraft wasn't functioning right and that maybe we'd lost it."[1]

For fourteen long, troubled hours, flight controllers tried to figure out what had gone wrong. They finally discovered that a faulty computer command had aimed *Magellan*'s radar into space instead of at Venus. Once corrected, the probe began sending back so much data that the computers at JPL had trouble processing all of it!

Until the early 1960s, scientists believed that the clouds of Venus were made of water vapor and that they surrounded a planet covered with swamps. Others thought that if swamps covered Venus, there must be life there as well. Venus is about the same size as Earth, 7,520 miles in diameter compared to our 7,926 miles. It is our nearest neighbor in the solar system. Scientists on Earth were eager to learn more about our "sister" planet. No matter how hard they tried, though, telescopes could not see through the cloud cover. It would take a

Venus (left) and Earth are called sister planets because they are nearly the same size.

space probe to tell us what was hiding under those thick clouds.

Mariner 2 was the first successful attempt by the United States to send a probe to another planet. It was launched in 1962 on a mission to fly by Venus. Although the probe observed Venus for only forty-two minutes, it measured very high temperatures in the planet's cloud tops. This surprised scientists. Other space probes were sent to Venus to try and solve the puzzle.

After looking at the data, scientists decided our nearest neighbor is not covered by swamps full of living things. Instead, Venus is a raging inferno. The planet has a surface temperature of nearly 900°F, which is hot enough to melt lead. Scientists wondered why Venus is so hot and what is under its thick blanket of clouds.

To help answer these questions, the spacecraft *Magellan* was sent on its way toward Venus in 1989. *Magellan* was the first space probe launched from a space shuttle. On May 4, 1989, the shuttle *Atlantis* released the probe from its payload bay. Once clear of the shuttle, an engine fired, sending *Magellan* on its way to Venus. The fifteen-foot-long, $900-million probe traveled through space for fifteen months.

After arriving at Venus on August 10, 1990, *Magellan* went into orbit around the planet. Once ground controllers had solved the radar problem, *Magellan* went to work. The probe's radar sent thousands of pulses per second through the thick clouds, toward the surface of

The space shuttle Atlantis *gets ready to launch the* Magellan *probe. The open payload bay of the shuttle is shown at the bottom of the photo.*

Venus. The radar echoes were bounced back to *Magellan* and then relayed to Earth. They were then turned into images by computers. At JPL, Stephen Saunders gathered hundreds of researchers to help examine the information. He asked the scientists to bring their graduate students to help with the huge project. Saunders explained, "There are students not born yet who will use this data and come up with things that I never dreamed of."[2]

The images showed a picture of a dusty, dry planet cracked and torn apart by thousands of large and small volcanoes. Some time in its past, boiling lava cut creases in the surface of Venus and pooled into hardened lava lakes and rivers. With the images sent from *Magellan*, scientists were able to examine areas on the planet's surface as small as a football field. According to *Magellan* radar expert James Head, the team feels "an extremely high level of excitement because of the level of geological detail we're seeing. . . . It makes your head swim."[3]

In addition to radar images of the surface of Venus, *Magellan* sent back data about the planet's atmosphere and chemical makeup. The surface pressure on Venus is ninety times greater than the pressure we feel on Earth. The planet's crushing atmosphere is 96 percent carbon dioxide. Carbon dioxide allows the Sun's light through to the planet's surface but prevents the heat from escaping back into space. Heat is trapped on the surface,

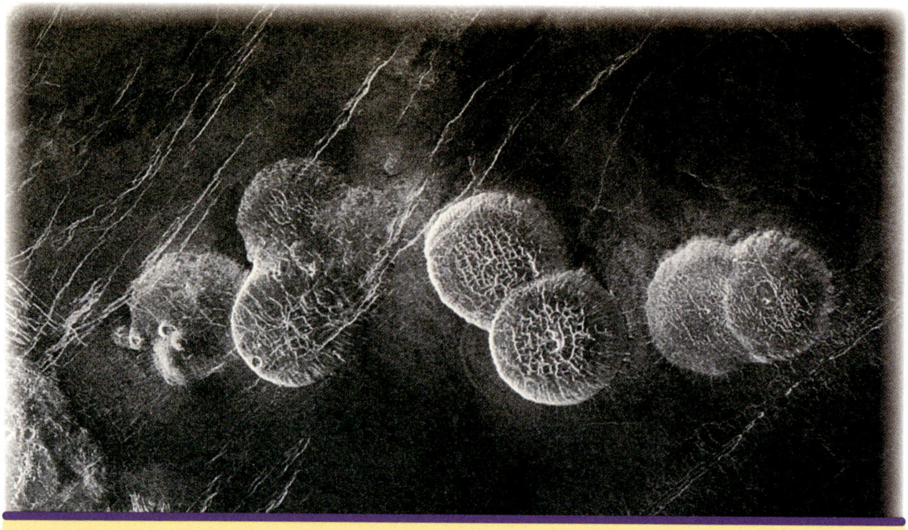

Magellan's radar mapped Venus's surface and revealed lava domes. They are called "pancake domes" because of their shape.

making Venus the hottest planet in the solar system. Enclosing this inferno are yellow clouds of sulfuric acid in the planet's upper atmosphere.

After making 15,034 orbits of Venus, *Magellan* had returned enough data to keep scientists busy for decades. The spacecraft mapped in detail 98 percent of the planet's surface during its four-year mission. With its funding gone, it was time for *Magellan* to get one last close-up look at Venus. On October 12, 1994, controllers at JPL fired five separate rocket bursts to slow *Magellan*'s speed and send it into lower and lower orbits.

A final command from JPL sent *Magellan* toward the blazing surface of Venus. *Magellan* program manager Jim Neuman reported, "It was interesting in that we had

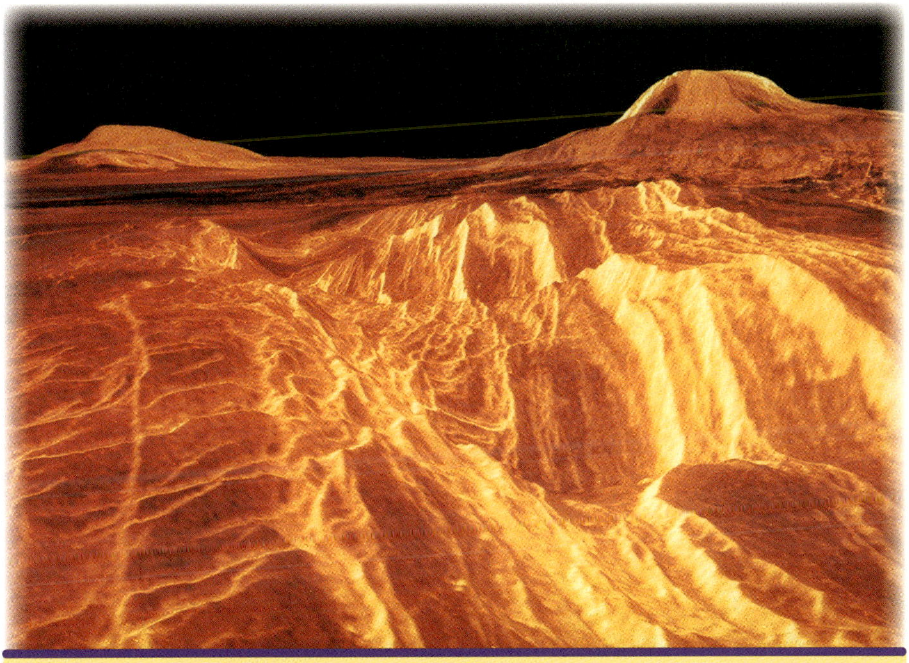

Gula Mons is a volcano on Venus. Magellan's *radar data was combined with color images from previous Soviet Venera spacecraft missions to create this picture.*

spent the entire night trying to kill the spacecraft. . . . It wouldn't lose attitude control, or memory, or engines. But then when it finally went in, the mood changed, and it was very emotional. I'll never forget it."[4]

Magellan will not be the last American probe to visit Venus. A proposed mission of the *Venus Sounder for Planetary Exploration* (*VESPER*) is being considered by NASA engineers. If *VESPER* becomes a reality, it will orbit Venus early in this century and record data about the planet's weather.

3

Voyager Flies by Saturn

According to NASA scientist Rita Beebe, "Saturn *is* astronomy to many people. Through even a small telescope, it's the most beautiful thing in the sky."[1] Scientists got a chance to really learn about Saturn, its rings, and its moons from probes called *Voyager 1* and *Voyager 2*. These two spacecraft were launched sixteen days apart in 1977, from Cape Canaveral, Florida.

The Voyager probes are about the size of a small car and look like big satellite dishes. They are powered by plutonium and carry many cameras and scientific instruments. Neither probe was scheduled to land on or even orbit the planets they visited. They were programmed only to fly by their targets, taking pictures and collecting data as they passed. They could not land on Saturn even if scientists wanted them to, because Saturn is made up of gas and has no solid ground. Along with Jupiter, Uranus, and Neptune, Saturn is one of the four gas giants in our solar system.

Voyager 1 reached Saturn in November 1980, followed by *Voyager 2* nine months later. Slightly different paths through space caused the delay in *Voyager 2*'s arrival. As the probes began sending back images of Saturn, a surprise was in store for the scientists. When viewed through telescopes on Earth, Saturn appears to have three large rings. These rings seem to be separated by darker bands that look like empty spaces. Pictures from the Voyager probes showed something very different. Bradford Smith, head of the Voyager imaging team, said,

The plutonium-powered Voyager probes flew by Saturn to take pictures and collect data.

"Saturn stunned us. What looked from Earth like empty gaps in the rings were filled with little rings."[2]

It turns out that Saturn has thousands of rings. They begin a little more than 4,000 miles above the planet's cloud tops and swirl outward for about 250,000 miles. Although Saturn's rings are very wide, they are only from 33 to 330 feet thick.[3] The rings are made up of chunks of ice and rocks. Some of the pieces are as small as a grain of sand, and others are as big as a bus.

During their flybys, the Voyager probes discovered eight more moons around Saturn, for a total of eighteen. *Voyager 1* collected data about Titan, the second largest moon in the solar system after Jupiter's Ganymede. Titan is bigger than the planet Mercury. It is made of solid material instead of gas, as are all of the moons. Titan is the only moon that is known to have a significant atmosphere. It is made up mostly of nitrogen. Since thick orange clouds completely cover Titan, photographs taken by the Voyager spacecraft were not able to show any details on the moon's surface.

Mimas is another of Saturn's moons that was examined by Voyager. Pictures showed that one third of the surface of this moon is a huge crater. An impact that created a crater of this size should have blown the moon apart. Somehow Mimas survived the hit. There are many other craters dotting Mimas's icy surface. *National Geographic* writer Rick Gore explained that "walking on Mimas would be a little like walking on a

From Earth, it looks as though Saturn has three large rings. However, Magellan *discovered that Saturn has thousands of rings (inset). The photo's color has been enhanced.*

Mimas, one of Saturn's moons, has an icy surface covered with craters.

large snow cone, with many ice chunks, some larger than a house, sticking up from the rubble."[4]

When *Voyager 1* and *2* finished their survey of Saturn and its moons, their trip was far from over. *Voyager 1* headed in a direction that would eventually take it out of the solar system. *Voyager 2* continued on its journey of discovery, reaching Uranus in 1986 and Neptune in 1989. The probes returned a huge amount of information to Earth, along with thousands of images of Jupiter, Saturn, Uranus, and Neptune. And they are still going.

As the twenty-first century begins, the Voyager spacecraft are at the edge of the solar system. They are still sending data back to Earth and are expected to continue operating until about 2020. Both spacecraft are expected to pass through a boundary called the heliopause that marks the end of the solar system. Reaching the heliopause will be a milestone in space exploration because no spacecraft has gone that far before.

Exploration of Saturn did not end with the Voyager missions. *Cassini* was launched on October 15, 1997. It is expected to reach Saturn on July 1, 2004. *Cassini* is the size of a small bus and is full of equipment to map

Voyager 2 *continued on to Neptune. This artist's conception shows the spacecraft with Neptune and its moon Triton.*

Cassini and its probe, Huygens, are joined at the Kennedy Space Center. They blasted off on October 15, 1997, and are scheduled to reach Saturn in 2004.

Saturn and its rings. Four months after its arrival, *Cassini* will release a probe called *Huygens* to explore Saturn's largest moon, Titan.

Huygens will hit Titan's atmosphere going 12,000 miles per hour. A parachute will deploy to slow the probe. On its two-hour ride to the surface of Titan, *Huygens* will measure the moon's wind speeds, gases, and temperature. The probe will then crash-land on Titan's surface, sending back pictures until it can no longer operate. Among other things, scientists will be trying to find out if early life exists on Titan. The moon's atmosphere is much like the one Earth had millions of years ago. Maybe Titan's orange clouds are hiding an amazing secret!

4

Galileo Explores Jupiter

"Oh, this is great!" said NASA scientist David Morrison.

"We now have confidence that we really are there, in the atmosphere of Jupiter, where nobody's ever been before. We've never probed into the atmosphere of a giant planet before. We've only looked down from above."[1]

As a 750-pound space probe raced through Jupiter's clouds, scientists cheered. The *Galileo* mission had been full of near disasters, but it was finally sending back data from the biggest planet in the solar system.

The planet Jupiter is big enough to hold more than one thousand Earths. If all of the other planets and moons in the solar system were glued together into a big ball, they would fit inside Jupiter with room to spare.

Jupiter, along with Saturn, Neptune, and Uranus, is one of the gas giants.

Jupiter is not solid like the Earth. The planet's small, solid core is surrounded by liquid hydrogen and then enveloped by miles of gases made mostly of hydrogen and helium. Jupiter's huge size creates enough gravity to keep the gases from escaping into space. South of Jupiter's equator, there is a swirling oval in the gas called the Great Red Spot. It is a gigantic, hurricane-like storm that is twice as big as Earth. The Great Red Spot has been visible since the seventeenth century, when

The planet Jupiter is not solid, like Earth. It is made of gases. However, Jupiter (left) is much bigger than Earth (right). All the other planets could fit inside Jupiter with room to spare.

people first used telescopes. It was not until the 1970s, though, that a spacecraft was launched to visit the gas giant.

Pioneer 10 flew within 81,000 miles of Jupiter in 1973 and sent back 500 images along with atmospheric data. In 1974, *Pioneer 11* came within 26,703 miles of the planet and took pictures of the Great Red Spot and the north and south poles. During the late 1970s, *Voyager 1* and *2* passed by Jupiter on their way to Saturn. They sent back 33,000 images of the planet and its moons. Each of these missions was a flyby. No spacecraft had actually orbited Jupiter. Then, in 1989, all of that changed.

The space shuttle *Atlantis* roared off the launchpad on October 18, 1989. Onboard was the *Galileo* spacecraft. Astronauts on the shuttle released *Galileo* into space. After it was a safe distance from the shuttle, *Galileo*'s rocket booster sent it on its way to Jupiter.

During its six-year trip to Jupiter, *Galileo* had many problems. In 1991, the umbrella-like main antenna on the spacecraft got stuck in a half-open position. Nothing ground controllers did opened the antenna fully. The *Galileo* team had to work around the problem and try to get the best results with the defective antenna. *Galileo*'s cameras worked perfectly, though. They were focused on the planet when comet Shoemaker-Levy 9 slammed into Jupiter's gaseous atmosphere. As some of the comet's fragments hit, fireballs shot 2,000 miles above the top of Jupiter's clouds.

Technicians at the Kennedy Space Center assemble the 750-pound Galileo *probe and its aeroshell.*

On July 12, 1995, nearly five months before arriving at Jupiter, *Galileo* released a 750-pound probe that raced toward the planet's atmosphere at 106,000 miles per hour. A parachute on the probe slowed its speed, allowing it to drift through Jupiter's clouds for an hour. There was no camera aboard the probe, but it measured wind speeds and the chemical composition of Jupiter's atmosphere before being destroyed by high temperatures and crushing pressure.

Finally, a little more than six years after its launch, *Galileo* arrived at Jupiter on December 7, 1995. Project scientist Torrence Johnson said, "We've had some hiccups and hard times here in the last few months. But the engineering and science teams have worked in a marvelous manner to get us ready for our encounter. We're ready to get into orbit."[2]

Once in orbit, *Galileo* began a two-year survey of Jupiter. Among the images sent back to Earth were pictures of giant thunderheads with lightning bolts a hundred times more powerful than any on Earth. There were also pictures of the nightside of Jupiter, glowing

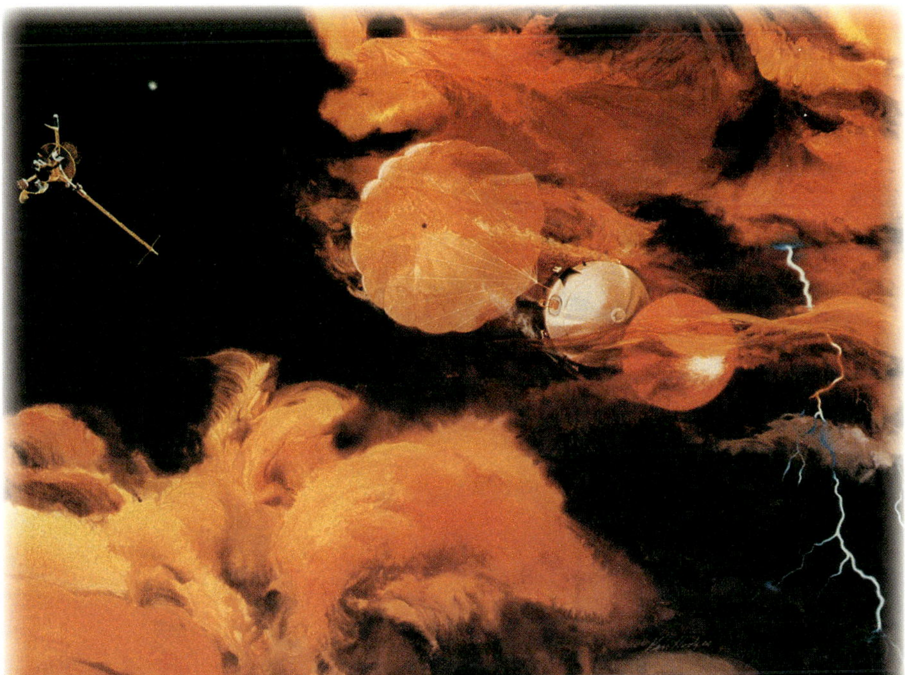

An artist's rendering of the Galileo *probe shows its descent into Jupiter's atmosphere. The parachute is fully deployed and the heat shield has been released.*

under the full moon of Io. Johnson said, "Jupiter dished up some juicy new tidbits we hadn't planned on."[3]

Galileo's looping orbits took it past Ganymede, Jupiter's largest moon and the biggest moon in the solar system. After looking at the pictures of Ganymede, space scientist James Head exclaimed, "It's absolutely stunning in its detail. It's an amazing place."[4] Instruments aboard *Galileo* discovered that Ganymede has a magnetic field, the first ever discovered on a moon in the solar system. "We're going to learn a lot about Ganymede as we try to understand this magnetic field," said *Galileo* magnetometer specialist Margaret Kivelson.[5] Scientists wonder what is happening inside Ganymede to create such a field.

Galileo also flew by Io, the most volcanically active moon in the solar system. Dozens of volcanoes dot Io's landscape, spewing lava onto its surface. Its crust is mottled with red, orange, and yellow areas, making it look like a giant pizza. Io's interior boils because it orbits so close to Jupiter's strong gravitational field. Constant pulling on the moon generates friction and heat in its interior, causing lava to spew outward. While *Galileo* was orbiting Jupiter, one of the volcanoes on Io erupted, spreading lava over an area as big as Arizona.

The moon that interests scientists the most is Europa. It is covered with a layer of cracked ice and towering icebergs that may cover a salty ocean. *Galileo* imaging team member Ronald Greeley said that "Europa

This is a false-color image of a volcanic feature on Jupiter's moon Io. It is named Amirani-Maui. At more than 160 miles long, it is the longest active lava flow known to exist in the solar system.

really is the gem of the solar system."[6] Europa may be the only place where we can find extraterrestrial life in an ocean. Scientists definitely want to take a better look at Europa to try and learn if a liquid ocean, and maybe life, exists on the moon.

One mystery surrounding Jupiter was solved during the *Galileo* mission. In the 1970s, one of the Voyager probes detected faint rings around Jupiter. The source of the rings was debated for years, until images from *Galileo* helped solve the puzzle in 1996. *Galileo* team member Maureen Ockert-Bell explained, "We can see the rings only when we're behind Jupiter, looking back at the Sun. It's kind of like when you're in a movie theater, and you turn around to look at the projector and hold up your hand so that the light's not in your eyes. Suddenly, you see a lot of tiny dust particles floating around."[7]

The faint rings floating around Jupiter in three distinct bands come from four small moons orbiting close to the planet's surface. Metis, Adrastea, Amalthea, and Thebe are constantly pelted with particles from space. The particles hit the moons at a high speed, kicking up dust. "It's like shooting a bullet into sand: the sand bursts up, and because there's little gravity, it escapes into the ring," said Ockert-Bell.[8] Gradually, the

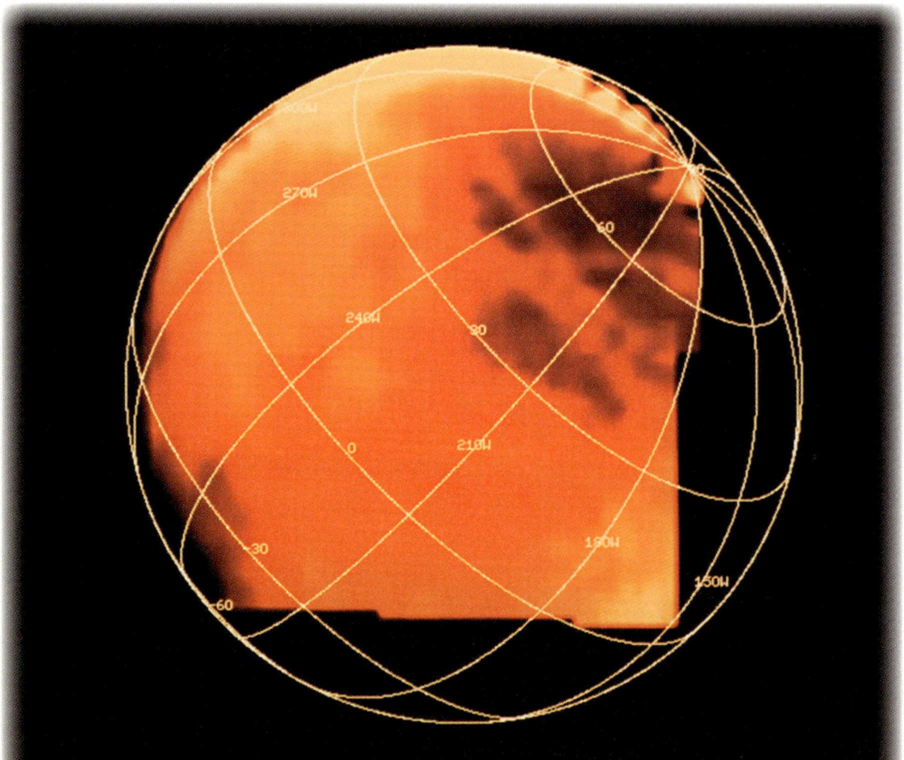

The frozen sulfuric acid on Jupiter's moon Europa is shown as the brightest yellow areas in this picture. Sulfuric acid is found in battery acid and in Earth's acid rain.

dust is pulled down to the surface of Jupiter, but it is replaced by new dust so that the rings are continually replenished.

Because the spacecraft was performing so well, the *Galileo* mission was extended. It was renamed the *Galileo Europa* mission because its main goal was to get a closer look at Europa. During several flybys of the moon, the probe sent back amazing pictures of objects as small as a truck. It appears that some of the icebergs on Europa move around like they are being carried on a current of liquid water. *Galileo* imaging team member Ronald Greeley explained, "These blocks of ice appear similar to those seen on Earth's polar seas during springtime thaws."[9]

With the help of additional probes, we will learn even more about Jupiter, and especially Europa. The *Europa Orbiter,* which has a proposed launch date of 2003, is set to arrive at Europa in 2007. As it orbits the moon, the spacecraft will bounce radio waves off of its surface. The return signals will help scientists decide whether or not there is a liquid ocean beneath the ice of this moon of Jupiter. If there is water on Europa, primitive life may also exist.

5

Pathfinder Lands on Mars

At JPL in California, scientists and mission controllers held their breath as they waited for the next maneuver. On the surface of another planet, 119 million miles away, three "petals" on the *Pathfinder Lander* opened, revealing a twenty-two-pound rover called *Sojourner* nestled within. Right on schedule, a camera aboard *Sojourner* whirred into action and began photographing the six-wheeled rover's slow trip down a ramp and onto Martian soil. As the first images from *Sojourner* began arriving at JPL on July 4, 1997, scientists cheered and clapped while the Beatles' song "Twist and Shout" blasted from Mission Control loudspeakers. For the first time in twenty-one years, a spacecraft from Earth was on the surface of Mars!

Mars, more than any other planet, has sparked the imagination of people living on Earth. In 1878, Giovanni Schiaparelli reported that he saw *canali* on Mars. The word *"canali"* was interpreted to mean man-made waterways instead of naturally occurring riverbeds. Could there be living beings on Mars, building canals? Science fiction writers loved the idea and began creating stories about little green men from the red planet. Movie producers jumped in and started making horror films about alien invasions.

When the Sojourner *rover began its trip down a ramp of the* Pathfinder Lander *onto the surface of Mars, it started taking pictures of the Martian soil.*

All of the speculation about intelligent life on Mars came to an end during the 1960s and 1970s. Four Mariner probes and two Viking orbiters with landers revealed a barren planet covered with red dust. Instead of finding a planet full of waterways, the spacecraft detected no liquid water or evidence of life on Mars. In addition, the probes analyzed the Martian air and found it was mostly carbon dioxide, which could not support life as we know it.

Even without liquid water or air that humans can breathe, Mars is still interesting to scientists. Other than the moon, Mars is the only place in our solar system that might support a colony at some point in the future. The temperature at the Martian south pole falls to nearly –200° F during the winter, while highs of 70° F are common at the equator in the summer. Even though the temperatures on Mars are colder than those on Earth, they are not too extreme to support a human colony. Mars is also fairly close to Earth. It would take six months of space travel to reach Mars, but a voyage of that length is possible.

In order to learn more about the red planet, NASA launched the *Mars Observer* in 1992. Unfortunately contact was lost with the spacecraft just as it was to enter orbit around Mars in 1993. At about the same time, budget cuts at NASA kept the space agency from planning any additional expensive projects. If NASA

Mars is interesting to scientists because it is the only other planet in the solar system that might someday support human life.

wanted to continue exploring space, it would have to cut costs and plan cheaper ways to fly.

Mars Pathfinder, which was launched on December 4, 1996, proved to be just such a project. The total cost of the mission was $265 million, compared to a $3-billion price tag for the Viking probes to Venus during the 1970s.

Made with a combination of new technology and a few spare parts, *Pathfinder* operated flawlessly on its seven-month journey to Mars. On reaching its destination, the spacecraft went into orbit around the red planet. It sent the *Pathfinder Lander* rocketing toward the Martian surface. Once its work was finished, the empty mother ship then crash-landed at another site, well away from the lander.

Two minutes before touchdown, a parachute on the lander opened to slow the descent. Then, eight seconds before striking the surface of Mars, airbags deployed and completely surrounded the lander in a cocoon of air. The *Pathfinder*

On December 4, 1996, Mars Pathfinder *blasted off and began its journey to Mars.*

Technicians run an airbag inflation test prior to the Mars Pathfinder *mission. These airbags softened the landing for the* Pathfinder Lander.

Lander hit the Martian surface at a speed of 23.5 miles per hour. It then bounced sixteen times over a half-mile area before coming to a stop. Once on the ground, the airbags deflated and moved out of the way.

After *Pathfinder Lander* opened, *Sojourner* made its way down the ramp and onto Martian soil. *Sojourner*, which looks like a toy built with an Erector set, got right to work taking pictures and analyzing the rocks and soil. Scientists at JPL gave the Martian rocks that *Sojourner* examined names such as Scooby-Doo, Yogi Bear, and

A parachute on the Pathfinder Lander *slowed its descent to Mars. Then airbags surrounded the lander just before it hit the surface. After bouncing to a stop, the* Pathfinder Lander *opened and* Sojourner *rolled out.*

Barnacle Bill. Two days after the rover began its exploration, chief NASA project scientist Matthew Golombek said, "We've got a perfect landing site, perfect instruments, and a perfect rover. We're all more excited than we could possibly believe."[1]

Some of the data sent back to Earth from *Pathfinder* and *Sojourner* showed that Mars had a major change in its climate between 2 billion and 3 billion years ago. "There have been two different climates on Mars," according to Golombek. "We don't know what triggered this climate change. It's a big question."[2] Thinking about the possibility that early Mars might have had

Sojourner *and* Pathfinder *underwent testing before they were launched to Mars.*

water, Golombek added, "Could life have developed on Mars if liquid water was stable? If it did, what happened to it? If it didn't, then why not?"[3]

While the solar-powered rover cruised around the Martian surface, images and reports about the mission were made available through the Internet on Earth. Children and adults around the world followed the journey of *Sojourner* from their computers. NASA counted more than 700 million visits to the *Mars Pathfinder* Internet site during the mission.

On September 27, 1997, signals to Earth stopped arriving from *Pathfinder* and *Sojourner*. The solar batteries on *Pathfinder* would no longer hold a charge. Since *Pathfinder* relayed communications from *Sojourner*, contact with the rover was also lost. Mission manager Richard Cook expressed the team's feelings when he said, "I guess this is a reluctant good-bye."[4] With the batteries used up, the lander and rover remain motionless and silent on the Martian surface. *Sojourner* was designed to function for only seven days and the *Pathfinder Lander* for only thirty days, but they both returned data for eighty-five days. Between them, the lander and rover sent more than 16,000 images back to Earth.

Just as signals stopped arriving from *Pathfinder*, another Martian probe arrived in Mars orbit. The *Mars Global Surveyor* did not land on Mars but began a four-year orbital mapping mission, sometimes getting as close as seventy-five miles from the planet's surface. *Global Surveyor* has sent back images of a world filled with wind storms, constantly shifting sand dunes, and huge dust devils. Some of the rotating columns of dust are five miles high. Mars team member Dr. Michael Malin said, "We see everything from sand dunes covered with frost, water ice clouds hanging over volcanoes, and ancient craters being eroded by high winds."[5]

For the next several years, an orbiter and lander is scheduled to blast off for Mars every twenty-six months. Among them is a mission that will use rovers to collect

The Mars Global Surveyor *has a camera (large silver cylinder at top) to photograph Mars.*

samples of Martian rocks and soil and return them to Earth around 2008. Human exploration of the planet may be a possibility by 2020.

When ancient people gazed up at the night sky, they wondered about the bright lights that looked down on them from overhead. In our world today, we have the ability to explore a few of these planetary objects and begin to solve the mysteries of the solar system. With the help of NASA's space probes and the Hubble Space Telescope, we have become familiar with some of the wonders of space. There are many exploratory space missions planned for the next century. Hopefully they will increase our knowledge and understanding of the universe in new and wonderful ways.

CHAPTER NOTES

Chapter 2. Magellan Orbits Venus

1. Kelly Beatty, "Working *Magellan*'s Magic," *Sky & Telescope*, August 1993, p. 16.

2. Ibid.

3. David Chandler, "*Magellan* Casting Venus in New Light," *Boston Globe*, September 24, 1990, Science and Technology section, p. 6.

4. Gregory Pope, "*Magellan*'s Last Voyage," *Popular Mechanics*, January 1995, p. 32.

Chapter 3. Voyager Flies by Saturn

1. Rick Gore, "*Voyager 1* at Saturn: Riddle of the Rings," *National Geographic*, July 1981, p. 5.

2. Bradford Smith, "Voyage of the Century," *National Geographic*, August 1990, p. 54.

3. Thomas Watters, *Planets: A Smithsonian Guide* (New York: Macmillan, 1995), p. 149.

4. Gore, p. 14.

Chapter 4. Galileo Explores Jupiter

1. Jane Kay, "Probe Phones Home from Jupiter," *San Francisco Examiner*, December 8, 1995, p. A2.

2. Earl Lane, "By Jove! As *Galileo* Passes Jupiter, a Probe Into the Planet's Turbulent Atmosphere Will Send Back New Scientific Information," *Newsday*, November 11, 1995, p. B27.

3. Kathy Sawyer, "*Galileo* Mission's Perseverance Extends Studies of Jupiter's Moons," *Washington Post*, December 26, 1997, p. A3.

4. Robert Cooke, "Mission Redeemed," *Newsday*, July 23, 1997, p. B17.

5. Ibid.

6. Sawyer, p. A3.

7. "Attack of the Flying Meteoroids!" *Time For Kids*, September 25, 1998, p. 4.

8. Dennis Cauchon, "Scientists Solve Mystery of Jupiter's Rings," *USA Today*, September 16, 1998, p. A3.

9. Richard Talcott, "Jumping Jupiter," *Astronomy*, June 1998, p. 40.

Chapter 5. Pathfinder Lands on Mars

1. Andrew Chaikin, "Back at Mars," *Popular Science*, October 1997, p. 64.

2. Robert Boyd, "*Pathfinder* Data Reveal Mars' Warm, Watery Past," *Houston Chronicle*, June 30, 1998, p. A5.

3. Mark Carreau, "NASA's Mars Team Notes *Pathfinder*'s End, Lauds Its Mission," *Houston Chronicle*, November 5, 1997, p. A2.

4. Ibid.

5. "Stunning New Mars Snapshots Show a Diverse Planet," July 19, 1999, <http://www.jpl.nasa.gov> (July 20, 1999).

GLOSSARY

asteroid—A rocky body found in space, sometimes called a minor planet.

astronomer—A scientist who studies the universe.

atmosphere—The layer of gas that surrounds some planets and moons.

comet—A celestial body that travels in a huge elliptical orbit. When orbiting near the Sun, it develops a long tail that points away from the Sun.

climate—The typical weather conditions of a place.

density—A measurement of the amount of mass that is contained in a volume of space.

extraterrestrial—Originating from or existing outside Earth.

flyby—Flight past an object or place.

heliopause—An area at the edge of the solar system where interstellar space begins.

Hubble Space Telescope—An orbiting observatory equipped with a very powerful telescope, designed to view objects up to 13 billion light-years away.

inferno—A place having great heat or flames.

interstellar—Among the stars.

kilometer—Metric unit of measure equal to about five eighths of a mile.

magnetic field—The area of space surrounding a planet or other large body where its magnetic force can be detected.

magnetometer—An instrument for measuring magnetic fields.

mass—The total amount of material a body contains.

National Aeronautics and Space Administration (NASA)—The United States government agency that administers the country's space program.

payload bay—The area of the space shuttle where cargo is stored and satellites are launched.

planet—Any of the nine large bodies in our solar system that orbit the sun.

solar system—The Sun, its planets, and their moons.

space shuttle—The winged vehicle used by NASA for transporting astronauts and experiments to Earth orbit.

FURTHER READING

Books

Branley, Franklyn. *Venus: Magellan Explores Our Twin Planet*. New York: HarperCollins, 1994.

Bredeson, Carmen. *Our Space Program*. Brookfield, Conn.: Millbrook Press, 1999.

Cole, Michael D. *Galileo Spacecraft: Mission to Jupiter*. Springfield, N.J.: Enslow Publishers, Inc., 1999.

Hitzeroth, Deborah, and Sharon Heerboth. *Galileo Galilei.* San Diego, Calif.: Lucent Books, 1992.

Levy, David, and Gregory Bridges. *Stars & Planets*. Alexandria, Va.: Time-Life Books, 1996.

Radlauer, Ruth. *Voyagers 1 & 2: Robots in Space*. Chicago: Children's Press, 1987.

Skurzynski, Gloria. *Discover Mars*. Washington, D.C.: National Geographic Society, 1998.

Vogt, Gregory. *The Solar System: Facts and Exploration*. New York: Twenty-First Century Books, 1995.

Internet Addresses

The Association of Universities for Research in Astronomy, Inc. *Space Telescope Science Institute*. "Hubble Space Telescope Public Pictures." 1990–1999. <http://oposite.stsci.edu/pubinfo/pictures.html> (February 11, 2000).

Dunbar, Brian. *NASA Homepage*. "Search the NASA Web." September 13, 1999. <http://www.nasa.gov> (February 11, 2000).

Jet Propulsion Laboratory. *JPL Web Page*. "JPL Current Missions." n.d. <http://www.jpl.nasa.gov/missions> (February 11, 2000).

National Aeronautics and Space Administration. *Galileo Homepage*. "*Galileo* Journey to Jupiter." n.d. <http://galileo.ivv.nasa.gov/index.html> (February 11, 2000).

National Aeronautics and Space Administration. *SpaceLink*. n.d. <http://spacelink.msfc.nasa.gov> (February 11, 2000).

National Aeronautics and Space Administration. *Welcome to the Cassini Mission*. n.d. <http://www.jpl.nasa.gov/cassini> (February 11, 2000).

INDEX